Do-It-Yourself Science™

EXPERIMENTS
with
SOLIDS,
LIQUIDS,
and GASES

Zella Williams

PowerKiDS press™

New York

Published in 2007 by The Rosen Publishing Group, Inc.
29 East 21st Street, New York, NY 10010

First Edition

Editor: Joanne Randolph
Book Design: Greg Tucker and Ginny Chu
Photo Researcher: Sam Cha

Photo Credits: Cover © Dave Nagel/Getty Images; p. 1 © Corbis; pp. 6, 7, 8, 9, 12, 13, 16, 17, 18, 19, 20, 21 by Adriana Skura; pp. 10, 11, 14, 15 by Cindy Reiman.

Library of Congress Cataloging-in-Publication Data

Williams, Zella.
    Experiments with solids, liquids, and gases / Zella Williams. — 1st ed.
        p. cm. — (Do-It-Yourself Science)
    Includes index.
    ISBN-13: 978-1-4042-3658-5 (lib. bdg.)
    ISBN-10: 1-4042-3658-9 (lib. bdg.)
    1. Matter—Properties—Experiments—Juvenile literature. 2. Solids—Experiments—Juvenile literature.
3. Liquids—Experiments—Juvenile literature. 4. Gases—Experiments—Juvenile literature.   I. Title.
    QC173.36.W555 2007
    530.4078—dc22

                                                        2006025643

Manufactured in the United States of America

Contents

What's the Matter?

All matter on Earth is a solid, a liquid, or a gas. Solids have a set shape. They have a fixed **volume**, too. Liquids flow. They have no shape of their own. Instead they take on the shape of whatever they are **poured** into. Liquids do have a fixed volume, though. Gases have no set shape or volume. They will spread out to fill whatever space they are in. They can also be **squeezed** into very small spaces.

Here you can see water in each of the three states of matter.

Solids Under Pressure

Sometimes matter changes from one state to another. When solids are heated, most of them change into liquids. For example, if you warm an ice **cube**, it will melt into water. When most liquids are heated, they change into gases. If you heat water, it will change into steam. **Pressure** makes ice melt, too. Try this **experiment**.

You will need

- ice cube
- thin wire or wire tie without the paper

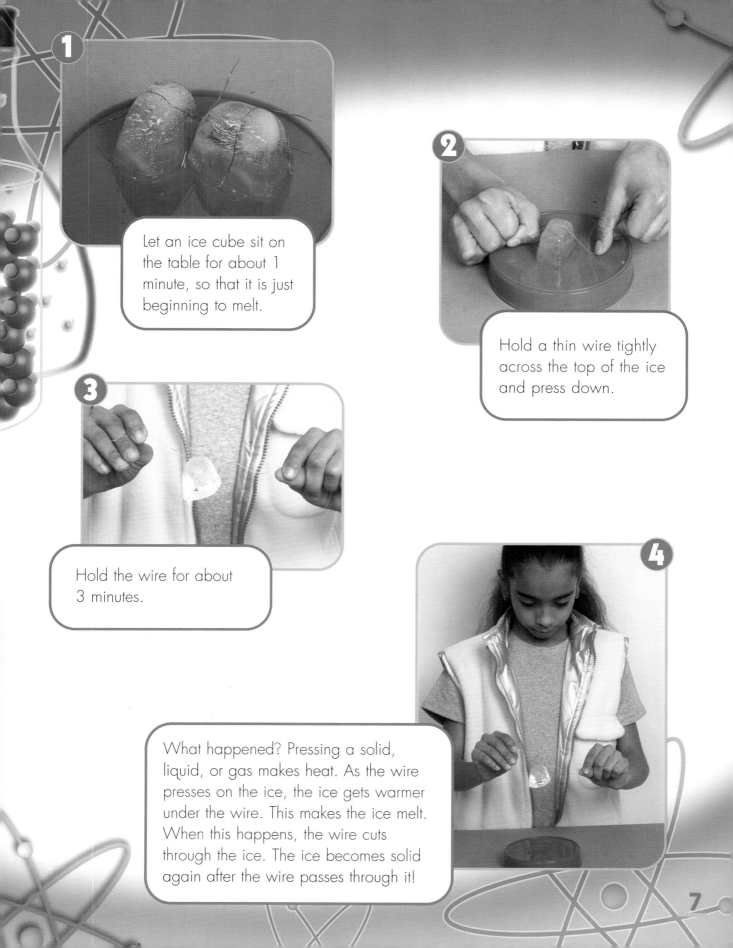

1 Let an ice cube sit on the table for about 1 minute, so that it is just beginning to melt.

2 Hold a thin wire tightly across the top of the ice and press down.

3 Hold the wire for about 3 minutes.

4 What happened? Pressing a solid, liquid, or gas makes heat. As the wire presses on the ice, the ice gets warmer under the wire. This makes the ice melt. When this happens, the wire cuts through the ice. The ice becomes solid again after the wire passes through it!

Solids Take Up Space

Solids have a set shape and volume. A solid's volume is how much space the solid takes up. Imagine you fill your bathtub all the way to the top. If you then get into the tub, the water will spill out. This is because your body takes up space in the tub. Try this experiment.

You will need

- plastic cup
- water
- two markers (different colors)
- a large piece of clay

1 Fill a cup halfway with water.

2 Use a marker to mark the **level** of the water on the outside of the cup. This mark shows how much space the water takes up.

3 Place a large piece of clay in the cup.

4 What happened? The level of the water changes because the clay takes up space, too. Mark the new level of the water with a different-colored marker. The difference in the level of water tells you the clay's volume.

5 Remove the clay from the water and form it into a new shape. Place the clay in the water again. The water rises to the same level as before. Why? Even though the clay looks different, its volume stays the same.

Making a Solution

Liquids can **dissolve** gases, other liquids, or even solids. It is called a **solution** when a gas, a liquid, or a solid dissolves in a liquid. Hot cocoa and ocean water are solutions. You may think matter that has dissolved in a liquid is gone, but it is not. Try this experiment.

You will need

- 2 tablespoons (30 ml) of salt
- 1/2 cup (118 ml) of warm water
- bowl

1 In a bowl that is not too deep, mix 2 tablespoons (30 ml) of salt into 1/2 cup (118 ml) of warm water.

2

3 Set your saltwater solution in a warm, dry place for a few days.

Mix the salt into the water until you cannot see it anymore. Is the salt gone?

4

What happened to your solution? Over time water evaporates into the air. "To evaporate" means "to turn into a gas." Solids do not evaporate in air. This means the salt is left behind in the bowl. It was there all along!

A Solid or a Liquid?

You know that solids have a fixed shape. You know that liquids flow and can be poured. So why would it be hard to tell whether something was a solid or a liquid? Here's an example of a **mixture** that acts like both a liquid and a solid. It's called ooblech.

You will need

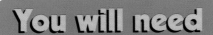

- box of cornstarch
- pitcher of water
- large bowl
- soup spoon

1

To make ooblech mix together a box of cornstarch and 1 1/2 cups (355 ml) of water in a large bowl.

2

Experiment with your ooblech. Does it pour? Slowly push your finger into the ooblech. It moves easily and acts like a liquid.

3

Slap the ooblech with your fist or squeeze it hard. The ooblech feels like a solid.

4

What happened? The cornstarch does not dissolve in water. If you move the ooblech slowly, the pieces of cornstarch slide around in the water. This makes the ooblech act like a liquid. Move the ooblech quickly, and the cornstarch gets pushed together. The water is then pushed out. This makes the ooblech seem solid.

Water, Water, Everywhere

Which liquid is everywhere you turn? The answer is water! People use water for bathing, drinking, and cooking. Plants need water to grow, and many plants and animals make their homes in lakes, rivers, and oceans. Did you know that most living things are made up mostly of water? Here's an experiment to try.

You will need

- scale
- carrot cut into several small pieces
- paper plate
- sheet of paper
- pencil

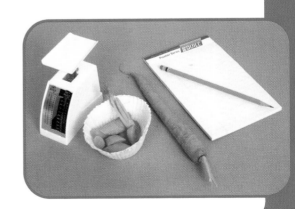

1 Use a **scale** to weigh a fresh carrot that has been cut into several small pieces. Record the weight of the carrot.

2 Set out the carrot pieces on a paper plate for several days. Then weigh the carrot again.

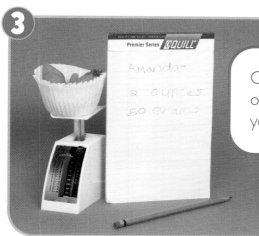

3 Compare the new weight of the carrot to the weight you recorded earlier.

4 What happened? The carrot weighs much less than it did before! That's because there is a lot of water in a carrot. The water evaporated when the carrot was left out in the air.

Air All Around

Gases are all around us. The air we breathe is a gas. We cannot see the air, though. So how do we know it is there? You can do this easy experiment to prove that air really is everywhere, even though we cannot see it. You are going to find out if an **empty** glass is really empty.

You will need

- glass
- paper towel
- sink or bowl full of water

1

Stuff a paper **towel** tightly into the bottom of a glass. Make sure that it stays there even when you turn the glass upside down.

2

Next fill a large bowl with water. Turn the glass upside down and push it quickly down into the water. Count to five.

3

Lift the glass out of the water without tipping it. Pull the paper towel out of the glass.

4

What happened? The paper in the glass stayed dry because the water could not rise up and get into the glass. Why? The glass was already full. It was full of air!

A Solid and a Liquid Make a Gas

Gases are one of the states of matter. When you blow up a balloon, your breath spreads out and pushes against the sides of the balloon. If you let the air out of the balloon, that air will spread out across the room. Here you will make a gas by mixing a solid and a liquid.

You will need

- funnel
- spoon
- baking soda
- balloon
- small bottle
- vinegar

1

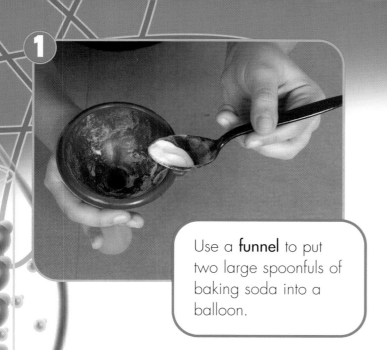

Use a **funnel** to put two large spoonfuls of baking soda into a balloon.

2

Fill a small bottle halfway with **vinegar**. Without letting any of the baking soda fall into the bottle, carefully pull the neck of the balloon over the bottle's opening.

3

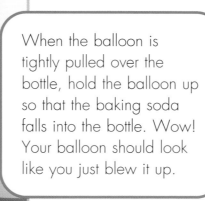

When the balloon is tightly pulled over the bottle, hold the balloon up so that the baking soda falls into the bottle. Wow! Your balloon should look like you just blew it up.

4

What happened? Baking soda is a solid and vinegar is a liquid. When these two things mix, they produce a gas called carbon dioxide. There is not enough room for the gas in the bottle, so it spreads out and fills the balloon.

19

Gas Far Apart, Gas Close Together

Solids and liquids cannot take up more or less space than they do. They have a fixed volume. Gas is different. Gas can spread out over a wide space. It is also possible to compress the tiny bits of gas into a small space. "To compress" means "to push closer together." Try this.

You will need

- plastic bottle with a cap
- water
- funnel
- food coloring

1 Find an empty plastic bottle and make sure the cap is on tight. Did you know that the bottle isn't really empty? It's full of air. Like all gases air can be compressed.

2 Squeeze the bottle to compress the air that is in it. The parts that make up the gas move very close together.

3 Fill the same bottle with colored water. Make sure the water fills the bottle all the way to the top before you put the cap back on. Squeeze the bottle.

4 What happened? The bottle full of water cannot be squeezed! The parts that make up a liquid cannot be compressed the way they can in a gas.

What in the World?

You have learned that everything in the world is made of matter. You have also tried some fun experiments to learn about the different kinds of matter. What do you know about solids, liquids, and gases now?

Look around you. Can you tell what kind of matter makes up each thing you see? By asking questions and doing experiments, you are learning about the world around you. What could be more fun than that?

Glossary

cube (KYOOB) A shape with six square sides.

dissolve (dih-ZOLV) To break down.

empty (EM-tee) Having nothing inside.

experiment (ek-SPER-uh-ment) A test done on something to learn more about it.

funnel (FUH-nul) A tool that is wide at the top and narrow at the bottom.

level (LEH-vul) How high something reaches.

mixture (MIKS-cher) A new thing that is made when two or more things are mixed together.

poured (PORD) Spilled or tipped out.

pressure (PREH-shur) A force that pushes on something.

scale (SKAYL) A tool used to tell how much something weighs.

solution (suh-LOO-shun) A mixture of two parts, one of which breaks down in the other.

squeezed (SKWEEZD) Forced together.

towel (TAU-ul) A cloth or paper for wiping or drying.

vinegar (VIH-nih-ger) A sour liquid used in cooking.

volume (VOL-yoom) The amount of space that matter takes up.

Index

Web Sites

Due to the changing nature of Internet links, PowerKids Press has developed an online list of Web sites related to the subject of this book. This site is updated regularly. Please use this link to access the list:
www.powerkidslinks.com/diysci/solids/